SAISAKETH KOPPU

The Evolution of Technology

Copyright © 2022 by Saisaketh Koppu

All rights reserved. No part of this publication may be reproduced, stored or transmitted in any form or by any means, electronic, mechanical, photocopying, recording, scanning, or otherwise without written permission from the publisher. It is illegal to copy this book, post it to a website, or distribute it by any other means without permission.

Saisaketh Koppu asserts the moral right to be identified as the author of this work.

Designations used by companies to distinguish their products are often claimed as trademarks. All brand names and product names used in this book and on its cover are trade names, service marks, trademarks and registered trademarks of their respective owners. The publishers and the book are not associated with any product or vendor mentioned in this book. None of the companies referenced within the book have endorsed the book.

First edition

ISBN: 979-8-9871765-1-1

*This book was professionally typeset on Reedsy.
Find out more at reedsy.com*

Contents

I   Part One

Chapter 1: Author Inspiration	3
Chapter 2: A Trip to The 1900s	5

II   Part Two

Chapter 3: The Computer	25
Chapter 4: Laptops	31

III   Part Three

Chapter 5: Smartphones - Part One	59
Chapter 6: Smartphones - Part Two	77
Chapter 7: Smartwatches	91

IV   Part Four

Chapter 8: Smart Home	105
Chapter 9: Autonomous Vehicles	116
Chapter 10: A Trip to the Future	123

Part One

Chapter 1: Author Inspiration

August 11th, 2017. This was the magical date when 24/7Tech, my technology YouTube channel, was born. After watching years and years of Internet personalities discuss their opinions about the latest technologies, it was my time to shine. As an aspiring 11-year-old, I created my channel and uploaded my first video about my computer at the time: the 2015 MacBook Air. By today's standards, that computer was prehistoric, featuring a 5th Generation dual-core processor and a measly 4GB of RAM. However at the time, it was a classy, sleek, fit in an envelope machine that suited all my needs. In the last 5 years of making articles, videos, and blogs, I've come to the realization of how fast we truly are progressing in the world of technology. It felt like just yesterday that the concept of an autonomous vehicle was in sci-fi movies, but people like Elon Musk is making this revolutionary technology a reality, day by day. What seemed like a computer that would last me a lifetime had evaded into the dust, a piece of prehistoric invention, one which seemed…

outdated?

Over 300 videos later, I felt a disconnect appearing with the current generation and the evolution of technology. A smartphone today is a sleek, metal and glass slab which can be so deceiving to truly understand the technology, complications, and hardware/software behind it. What better way to convey this evolution than a easy-to-follow book for you all to understand!

I would like to dedicate this book to my dad, Ramesh Koppu, for encouraging me in my technology pursuits and constantly keeping me on my feet, knowing what I am capable of. This book would not have been possible without the push and drive you have given me.

I would also like to dedicate this book to my mom and brother, Vasavi Maram and Saisatvik Koppu, respectively, for their support throughout my technology journey.

Chapter 2: A Trip to The 1900s

Lets take a trip back to the 1900s. Where did the technology we have truly come from, and what are the foundations of it?

The "computer" might be perceived today as a laptop, desktop, or PC. However, a computer is simply a piece of hardware that can be programmed to do particular tasks the user would like it to do. The key word in the description above is "programmed." The first programmable computer, called the "Z1" by German Konrad Zuse, could simply add and subtract, but it was programmed of course. This was the birth of the first computer, albeit with many more pins, plates, and moving parts than there are today. But for 1936, not too shabby.

The Evolution of Technology

Picture of a "Calculating Machine"

Now, a company still famous for its computers today, HP (Hewlett-Packard), was founded by Bill Hewlett and David Packard in 1939 in Palo Alto, California. Talk about consistency, still dominating the computer market near a century later! In the midst of this computing revolution, World War II had struck, and the race to develop the next best thing was on. In 1941, Konrad Zuse, the inventor of the Z1, created the "Z3," the world's first digital computer, albeit it was destroyed in Berlin raids during World War II. As an answer, the United States developed the first digital computer in the USA, the ABC (Atanasoff-Berry Computer).

However, these were simply budding stones for what was to come next. Introduce, the ENIAC. With over 40 cabinets, 6,000 switches, and 18,000 vacuum tubes taking up over 1,500 square

Chapter 2: A Trip to The 1900s

feet of space, the ENIAC was legit. The ENIAC set in place what we know today as the "Computer Age." Compared to the Z1 from before which could barely operate, this computer could compute *thousands of operations per second.* Now, today's computers can do much more, but this was still revolutionary at the time.

An ENIAC from the 1940s

Fast forward a couple of years, and the first computing language was finally created, and it was called *COBOL*. Created by Grace Hopper, who is known as the *"First Lady of Software,"* she truly set forward the foundation we have today for programming languages like C, C++, and Java. Created in 1953, COBOL is widely utilized even to this day in mainframe computers, and newer machines have moved past this programming language, but it still has a significant presence in this world.

The Evolution of Technology

Sure, we now had a way to program by using COBOL. What about the integrated chips that we use today all around our phones, tablets, and computers? In 1958, Jack Kilby and Robert Noyce introduced the world's first "chip" — not to be mistaken with the food, of course.

> Fun Fact: Jack Kilby was awarded the *Nobel Prize in Physics* for the development of the computer chip!

Until the year 1965, computers, machines, and anything of that sort were primarily limited to engineers in a lab environment. The Programma 101 changed that all. Designed in Italy, this was the first computer that the public could purchase. In all honesty, it actually looks pretty good too. This machine weighed around 65 pounds, had 37 clickable keys, and even a printer built into it! This was the first machine the normal —"normal" as in publicly available, not affordable — person could purchase, and it was capable of doing calculations like absolute value, square roots, and more. The Programma 101 is known widely as the *first personal computer*.

Chapter 2: A Trip to The 1900s

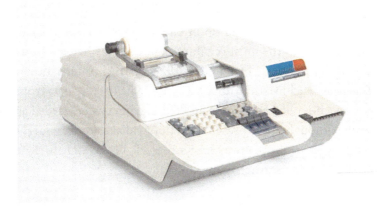

The Programma 101

MacOS. Windows. Linux. What are these? These are all *operating systems*, but first of all what is an operating system? An operating system is a piece of software that is loaded onto a computer to manage all of the applications, data, and programs on a computer. An operating system manages almost all of the tasks done by a computer. So, who pioneered the operating system, and what was the first one? In 1969, the world's first operating system, *Unix*, was born at Bell Labs. Unix was not only revolutionary and ahead of its time, the operating systems we use today are derivatives of it! If you own a computer utilizing MacOS (Apple Macs), or even a computer running Linux, these are both derivatives of the Unix operating system developed over 60 years ago. Crazy, right?

The Evolution of Technology

> Fun Fact: The founder of Bell Labs is the inventor of the telephone himself: Alexander Graham Bell!

We also haven't touched upon data storage yet, and this is because one of the most revolutionary data storage methods was the *floppy disk*. Invented in 1971 by IBM (International Business Machines), the floppy disk was able to store data from one machine, and be used on another machine when needed as well. Although in the present day, floppy disks are very rarely used, these set the pathway for more sleek, faster, and efficient storage methods to arrive into the industry.

An IBM-made floppy disk

Lets cycle back to personal computers, shall we. Now that we have storage mechanisms, operating systems, computing languages, and computer chips out of the way, when was a

Chapter 2: A Trip to The 1900s

"normal" computer made?! In the 1970s, personal computers became more and more popular. One computer released by Xerox has design language synonymous with today's computers, and in particular, Macs. The Xerox Alto, released in 1973, had a *606x608* display, and a keyboard and mouse! This computer has the design language of a traditional all-in-one computer we see today: a display, keyboard, and mouse, all in one package. However, the machine itself was known as "sub-par" by critics. So why was it revolutionary, one might ask? Well, Apple's early machines were actually based off of this design!

The Xerox Alto 1

The date was April Fools of 1976, and the computer-world was about to change forever. Steve Jobs and Steve Wozniak, two brilliant men, would invent a machine and company that would take over the world for the foreseeable future: Apple. Introduce, the Apple I. In all honesty, this was not the revolutionary product Apple created, but it was their first step in development. The Apple I was a computer which packed a single-circuit board and Read Only Memory (ROM), but the primary aspect of this computer was that users would have to add on vital parts of the computer, like power and a shell. For that reason, customers were not exactly thrilled about the computer, and it isn't the reason why Apple is where it is today.

The Apple I computer

The true computer which changed the world was the Apple II, introduced in 1977 by Steve Jobs and Wozniak. This computer was a full-fledged, 8-bit home computer which removed the aspect of user-supplying — creating an end-to-end product which users could use effortlessly. The graphical display used by the Apple II was second-to-none by any other computer on the market, and it was a highly successful product. By hiding things like wires, including a built-in keyboard, and having TV

Chapter 2: A Trip to The 1900s

compatibility, the Apple II was the pinnacle of a computer at the time, and it had packed features so beyond its time, propelling Apple into the stardom of computers.

> Fun Fact: Apple produced the Apple II until 1994! Through 17 years of existence, the Apple II became the face of computing.

The Apple II Computer

What about the competition? Surely other companies weren't just sitting back looking at Apple take over computer market, right? In 1981, IBM released their first personal computer called the *IBM PC (nickname: Acorn)*, to rival the Apple II. The IBM PC — which stands for "personal computer" — could process information much faster than the previous mainframe technology that IBM was putting out, and this was at the fingertips of consumers now as well. Having access to this level of technology and speed was not normal to IBM customers in such a small footprint, but it was now possible.

The IBM PC, circa 1981

In 1983, Apple introduced the *LISA*, which was one of the first computers ever to incorporate a Graphical User Interface (GUI), with even a drop down menu, icons, and more. Despite this innovation, it was not known as one of Apple's more successful commercial products because of what they released in 1984, and also because it costed $10,000. That is equivalent to nearly $27,000 in 2022 after accounting for inflation, demonstrating the lack of affordability, hence why it only sold 10,000 units.

Chapter 2: A Trip to The 1900s

Apple Lisa Computer

In other things that occurred in 1983, the world's first "laptop" was invented: the Gavilian SC. Featuring a "folding" screen, it truly wasn't the best of laptops, and doesn't really look like the clam shell-based laptops we use today, but it featured a display that folded, and isn't highly regarding as a mass-market laptop. Let's track back to Apple's Lisa. Following the commercial failure of the Lisa, Apple decided that it was time they provided a machine to the consumers which was elegant, sleek, and everyone would want. It was time for their biggest launch since the Apple II: the Macintosh, or "the Mac" for short.

The Macintosh was the revolutionary point in computing to introduce something called the *All-In-One.* This computer had the graphical user interface. It had the built-in keyboard

The Evolution of Technology

and mouse. It had the high-quality display (for the time). It included MacWrite, a word processor that Apple had developed, and didn't break the bank as much as the Lisa. It started at $2,500, making it available to many more people than other computers that Apple and other companies had created. The "Mac" moniker has also survived the years — as Apple continues to use the Mac moniker in their computers today.

The Apple Macintosh 128k, circa 1984

During this time of personal computers, laptops started to catch fire as well. The first mass-produced, commercially successful laptop was the *Toshiba T-1100*. It started at $1899, which was actually quite a bit cheaper than Apple's Macintosh, and had a folding display, and a built-in keyboard. For 1985, this "laptop" was revolutionary because of its low weight, portability, and ease-of-use. Weighing 9lb, this laptop was definitely not

Chapter 2: A Trip to The 1900s

as light as the ultrabooks we have in present day, but it was manageable to take places and use in multiple locations, which was something the computing industry hadn't seen before.

Toshiba T-1100

1985 also marked the invention of something *huge* in the computer world: Windows. In response to Apple's Lisa which featured a GUI, Microsoft felt the pressure to create their own operating system with a GUI. Windows, as most of us know today, is still widely utilized and is the most popular operating system in computers to this day. The birth of this operating system over 35 years ago demonstrates the evolution we have had from the beginning OS with a GUI, to a full-fledged operating system used all across the world.

A question you may be asking yourself is how were these

devices all connected, and how could you access particular things off of the Internet, and use them on your own computer. Well, in 1989, a British researcher named Tim Berners-Lee proposed an idea which eventually became the World Wide Web. Whenever you type in "www," this is referring to the world wide web, which was invented in 1989. He also created the foundations for the programming language *HTML,* which would be the programming language the web used.

The World Wide Web

In 1996, the phrase "Google It" was truly born. Larry Page and Sergey Brin at Stanford University created the Google search engine, and this has been by far the #1 search engine ever since. At the time, the success level of their search engine and algorithm to map users' inputted key words into search results was impeccable, and to say that it was born in a dorm

Chapter 2: A Trip to The 1900s

room is pretty neat.

But wait, where has Apple been since the release of the Macintosh in 1984? In 1985, Apple went through a *major* change in their leadership which sent them behind for the next decade or so. By removing the brains and master of design Steve Jobs, and replacing him with John Sculley, the new CEO, Apple's sense of design and innovation was abruptly halted. This decision was made because of Steve Jobs' lack of compatibility with his board of directors, and they voted to remove him, which would prove deadly to the company.

John Sculley (ex-Apple CEO)

So, was Steve Jobs just idle during this time? Well, nope. Jobs developed and opened a company called NeXT using $12,000,000 of his own money, and started to create computers with that company. Unfortunately, this company was not very successful, and it didn't gain the same fruition as the Apple products had gained in the past.

NeXT Computer developed by Steve Jobs

However, in 1997, everything had gone downhill for Apple. In the 1990s, Apple started to lose $1 Billion yearly, and this loss yearly had put the company in a very bad place. During the 1990s, Apple filed a lawsuit against Microsoft for copying their Operating System, and as reparations, Microsoft had invested $150M into Apple in 1997, alongside a plethora of other moves to make Apple the company it is today. For starters, the acquired NeXT and Steve Jobs in order to hopefully revive the company through Steve Jobs' brilliance, and he did exactly so with the iMac G3.

Chapter 2: A Trip to The 1900s

1998 iMac G3 pictured on the right

Part Two

Chapter 3: The Computer

Before we get into the 21st century, and how various types of computers have evolved recently, how does the computer really work? Let me first give you the basic definition of a computer, then break it down into more complicated parts. A computer is a machine that takes in information, processes the inputted information, and then outputs the information that it has created. Simple, right? Now, let's get a little more technical.

This "information" that the computer takes in is actually called *data,* which is raw information can be represented in numerous different ways. The process of taking in this data is called *input,* and the information it returns to the user is called *output.* Now, what happens if we want to simply store a piece of data to process later? This data would be stored in what's called the *memory* of a computer.

The Evolution of Technology

Inside of a Computer

Now that we know the main parts of a computer, what are some examples of these parts, and how do they play a part in the overall functioning of a computer?

- **Input:** This can be things like a *keyboard and mouse* to give information to the computer to start the processing. Any other sensors (light, heat, temperature) or even a microphone can be considered as input into the computer!
- **Memory:** Have you ever seen "GB" written on any phone, laptop, or desktop you've purchased? This is the amount of storage in the device, and there are many different types of storage available. Hard drives used to be the norm in large computers, as they can hold more information but are

Chapter 3: The Computer

slower. Solid State Drives (SSD) are much faster, yet more expensive and can hold less information. Most laptops and ultrabooks that you purchase will come with an SSD instead of Hard Drives.

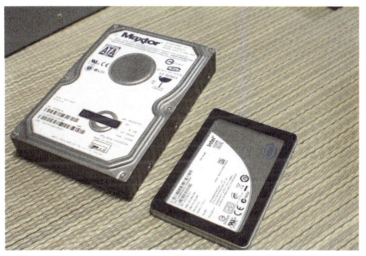

Hard Drive (pictured on the left) vs Solid State Drive (pictured on the right)

- **Processing:** This is where the magic happens inside of a computer. All of that data taken in, here is where the computer processes it. The Central Processing Unit of a computer, or the CPU, is the brain of the computer, where all processing occurs. All of the input taken in is processed rapidly (causing heat, which is why we have fans in our computers!), and then produced into output for the user.

- **Output:** So how does our computer give us our information after it has processed it? Well, there are many ways. Whether it be the primary way of showing it to us on the screen to view our work, listen to our audio files on the built-in speakers, a projector, or a connected printer, our computers have many ways to produce output for us.

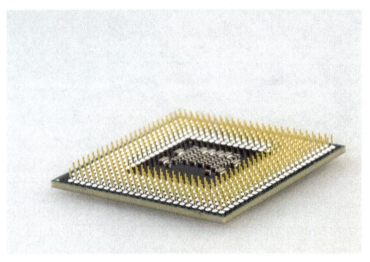

A Central Processing Unit (CPU)

Now that we have the overlying concepts out of the way, how does a user tweak a computer to make it solve the particular problems they might have? Well, the answer is a *computer program.* When the computer was first invented in the early 20th century, they weren't built to do the tasks they are today, instead just calculate simple math problems and operations. Today, we ask our computers to edit and produce our videos,

Chapter 3: The Computer

write our essays, make our music, and even web browse!

However, similarly to the computers of the stone age, all of the things our computers do today can also be translated into numbers, once broken down enough. Around 50 years ago, to write a letter to send to someone, one would have to first create their own program which allowed the computer to be able to take in the user's inputted text, process that text, and then output it back to the user. In most cases, these programs would be tedious to write, and took longer than writing the letter itself! After a while, computer programs called *word processors* were invented to make this task easier on everyone's hands, leaving out the need to create a new program every time you wanted to write something.

Microsoft Word (circa 2016): A Modern-Day Word Processor

Now that we've tackled the problem of running a singular program on a computer, what if I wanted to run multiple programs? Would you need to create a new program every time you wanted to perform a different activity? Well, until the advent of the *operating system*, that was exactly what you would have to do. However, the invention of the operating system made this whole process much less tedious on everyone's hands. Since there are so many similarities in all of these programs and tasks they all had to complete, operating systems simplified that entire process down. Examples of modern-day operating systems include MacOS, Windows, Linux, and more.

MacOS Big Sur Operating System

Now that the makeup of a computer makes more sense, let's get onto the fun stuff, shall we?

Chapter 4: Laptops

Let's start off with the first "laptop" ever invented. This computer was not the first modern foldable computer, but it was the first "modern laptop" that was ever made, and there were similar qualities of this laptop and the laptops of today's day and age. It featured a tiny screen, a plastic frame with chunky and clunky parts, and costed $8,150 on launch day. Adjusted for inflation, in 2022 this laptop would've costed over $20,000! For reference, a brand-new Toyota Corolla would cost around the same — expensive, right? 11 pounds and a 320 x 240 display, *"thick and ugly"* is what kick started this beautiful invention.

The Evolution of Technology

The Grid Compass - the first laptop

In the late 20th Century, IBM was one of the most influential and advanced computer companies, and they jumped on the laptop train early, with the PC Convertible. At first glance, this computer doesn't look and have a clam shell type design of the Grid Compass and laptops of today's generation, but it costed "only" $2,000, featured a 3.5" floppy disk, and a 10 hour battery! The laptops of this generation, and the PC Convertible in particular had a cool feature where the display can be detached from the computer itself, something we would see in future laptops as well.

Chapter 4: Laptops

IBM PC Convertible from 1986

1989 introduced the Apple Macintosh Portable, which is rendered as one of Apple's worst products ever. Suffering from the loss of Steve Jobs, the 1990s was primarily a period of time when Apple did not experience much success, and stagnated a lot. However, this laptop is still monumental in that it was Apple's first portable laptop, albeit its unappealing characteristics. The laptop featured a 9.8 inch, 640x400 display that was only *black and white.* Coming in a cream, non white color costing over $7,300, this laptop simply was too expensive and too under powered in order to be purchasable and mass-marketable for consumers. Yet, it was Apple's first "laptop".

The Evolution of Technology

1989's Apple Macintosh Portable

1991, however, introduced one of Apple's most famous computers ever made: the PowerBook. These machines finally resembled something close to a laptop of modern day, and it was released in three different models on launch day: the PowerBook 100, 140, and 170. Yes, higher numbers meant more expensive and more feature-packed models. This laptop was so powerful in the industry that it captured close to 40% of sales in the market, exemplifying its impact on the market and world. The PowerBook 170, the highest-end laptop of them all, even featured an active matrix display, displaying the modern features being packed into these smaller framed machines from Apple.

Chapter 4: Laptops

Apple PowerBook 100

Have you ever seen a computer meant for businessmen and workplaces, one designed for optimal output and not many flashy features? Are you thinking of a Lenovo ThinkPad? Well, well, well. The first ThinkPad was released in *1992*, making it over 30 years old now. The twist is, the ThinkPad, for you young readers, was actually owned by IBM at the beginning. Sold to Lenovo in 2005, the IBM ThinkPad was a commercially successful laptop known for it's slate black, boxed out design from day one. The design language truly speaks volumes on ThinkPad and how iconic it has became in the laptop market for simply reliability and productivity.

The Evolution of Technology

Fun Fact: The ThinkPad introduced *TrackPoint,* the little circle in the middle of your keyboard you could utilize as a mouse for your screen.

The first IBM ThinkPad

Let's fast forward a good amount, shall we? The century was about to end, the year was 1999, and Apple had dropped — drum roll please — the iBook. Coming in savvy, vibrant, and beautiful colors and marketed as the "iMac to go," the Steve Jobs-led company had brought a new, peppy look to the laptop industry that took it by storm, and was appealing the *normal consumer.* Unlike the ThinkPad, which was meant

Chapter 4: Laptops

for businessmen and company executives, this was for your average Joe, who might want a laptop to simply surf the Web on.

The iBook in an "average household"

The iBook came with a clunky design by today's standards, but the infusion of color made it so sought after. The iBook simply made things easier, and introduced the Apple laptops we know today: easy to use, functional, and reliable. This was the first Apple laptop to achieve Wi-Fi connectivity, a USB port, and a 12.1" display as well. It infused elements to attract the normal consumer, whether it be weighing "only" 6.6lb, having a built-in handle to carry it, or coming in at a starting point of $1,599. All of these features meant that Apple, of course, had to cut corners with the engineering of this computer. All in all, however, it

The Evolution of Technology

was still a great machine and an icon to enter the 21st century.

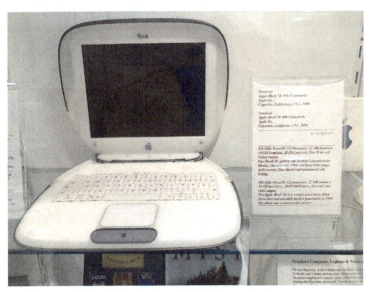

The rise of the 21st century meant that the laptop market was open for any worthy company to take over and rise up to dominance. The laptop market was, and still is, one that can be dominated only through variety, cheaper prices, and mass-production for high output. There was one company that went off to a racing start to dominate the industry: Dell. Dell has been a company that has dominated the mass-market laptop — and in the 2000s, every household wanted to get their hands on a laptop. With the introduction of thinner bezels, colors like silver, and a more boxed out look, Dell dominated this industry.

Chapter 4: Laptops

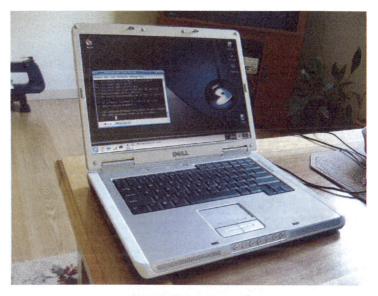

2007 Dell Inspiron 15"

We cannot move to arguably the most influential laptop launch of the 21st Century without discussing the state of IBM ThinkPad. In 2005, IBM decided that the laptop industry wasn't for them, and sold it off to Lenovo for them to manage the ThinkPad division. There's controversy regarding whether Lenovo brought the laptop line to new heights, or ruined its originality and design. General consensus, however, is that Lenovo blended in some modern aspects of laptops, while maintaining key ThinkPad features like ThinkPoint, the slate black boxed out design, and more.

2008 introduced a market breaker. This laptop arguably dictated the trend for the 2010s and 2020s in regards to design language, elegance, sleekness, and reliability: the MacBook

The Evolution of Technology

Air. Debuted as the thinnest laptop in the world, and pulled out of an envelope, Steve Jobs created a thing of beauty with the MacBook Air. It was incredibly minimalist, met with skepticism and doubt because of its lack of IO, older ports, and the omission of speed. However, this would be the staple laptop in aspect of design, sleekness, battery life, and simplicity.

Steve Jobs unveiling MacBook Air from an envelope

The MacBook Air actually launched at a price of $1,799, which was pretty pricey for the time. The laptop quickly reduced in price to $1,299 by 2010 as the industry and Apple's other laptops started to catch up. The MacBook Air had everything the tech-savvy, feature-seeking consumer would want. Thinnest laptop? Check. Less than 3 pounds? Check.

Chapter 4: Laptops

Glass track pad with MagSafe? Check. It hit every box and packed so many new features of Steve Job's creation that people were willing to overlook the fundamentals it had omitted (which its counterparts would soon do as well following the tech giant). This was also the last "blockbuster" laptop Steve Jobs would develop for Apple before his unfortunate death due to cancer in 2011.

One of the sleekest laptops of all time: the MacBook Air

Another piece of design language which has been influential since its release was the 2012 Microsoft Surface. This computer wasn't your traditional clam shell machine, instead it was a tablet with a keyboard and stand attached, making it one of the most versatile machines ever built. It was designed to fill multiple jobs in one computer, whether it be gaming on a tablet, writing emails on your attached keyboard, this was the do-it-all machine. However, it started at only $499 and you could spec it up to a more capable model for money. Albeit, the lowest end model came with an ARM processor that wasn't really even close to the Intel or AMD offerings of the time.

The 2012 Microsoft Surface

Let's track back to the professional, more capable machines, shall we? The next big breakthrough of Apple's laptops was the 2012 Retina MacBook Pro. This was *the* machine for quite some time, and is still known today as a more than capable machine of anything you throw at it. For 2012, a 720p display was still known as respectable, but this computer threw a 1440p, color accurate, thin-bezel display at consumers, and they loved it. This laptop first released in a 15" model, with a

Chapter 4: Laptops

13" following it up later on from Apple. However, these pro features came at a pro price tag, and this laptop started at a whopping $2,199, making it only for a select user, and not all. Despite the heavy price tag, this laptop set forward Apple's design language, displays, port selection, track pad, and so many more features of the future.

The 2012 Retina MacBook Pro

The return of Lenovo and the ThinkPad. This time around, focused a little more on personal and consumer use, while maintaining the loved features of ThinkPad and their loyal consumers. The X1 lineup, which Lenovo designed for personal use, consisted of numerous different laptops throughout the years — the X1 Carbon, X1 Yoga, X1 Tablet, and more. However, the first, most popular and groundbreaking models of these was the 2012 X1 Carbon.

This machine simply was known to do everything right, and nothing wrong. It omitted the flashy features and tricks of the

MacBook and Surface lineup, and went back to simply build a practical computer. With a 14" display, multiple RAM and SSD models to choose from, and a carbon fiber design, this computer knocked down the basics perfectly, and honestly did nothing much more. This was a computer aimed at not the most tech-savvy person, and the brand of ThinkPad elevated its media presence due to the automatic features that would be featured in it, just because of its name.

2012 Lenovo X1 Carbon

Following the success of the Surface, Microsoft decided to try its hand again, this time at a more premium version of the Surface with a couple more flashy features to it. This machine is the *Surface Pro*, a laptop many people love to this day. This 2-in-1 device started off at $899, and still only featured a 10.6" display, which might leave you wondering why it was so expensive for such a small computer.

It was the *panel* that made this laptop more expensive than before, as it featured a 1920 x 1080 display with more PPI than even the iPad at the time! It also featured a stylus, something we are used to seeing on newer Surface products as well, which worked well, but wasn't near the design language or accuracy of Apple's future Apple Pencil or even Microsoft's future Surface Pen. The keyboard was also **not** included with the purchase, which costed upwards of $100 to purchase additionally.

Microsoft's Surface Pro

The Evolution of Technology

This next computer might have revived this companies upscale laptop market, with it becoming one of the best computers in the next decade: the Dell XPS. The 2013 Dell XPS 13 wasn't really up to date with the design language and Apple still ruled that area, but it set the foundation for rapid and exponential improvement. Being a company that loves changing things all the time, and generally first to new hardware and software, Dell was able to accelerate off of the mediocre 2013 XPS in order to create a lineup of mass-marketable laptops.

As for this 2013 XPS, though, it was marketed as the Windows MacBook Air, as Dell aimed to create a sleek design at an "affordable" price point, starting at $999. Featuring the paltry Windows 8 software, it never really gained steam with consumers, despite its soft touch keyboard, thin and light design, and thinner than normal bezels on the display. Being touchless, it simply lacked the attributes of a Windows 8 machine, and never really meshed well with that software — similar to many other Windows laptops.

Chapter 4: Laptops

2013's Dell XPS 13

The most iconic Surface product of all time has to be the *Surface Book*. Introduced in 2015, this laptop was simply cutting edge in its features and versatility, and nothing has really came close ever since. The concept of being able to completely detach the display and use it alone with the industry-leading hinge has never been beaten, and for the right person, it was **the** product.

However, it was expensive, didn't really get some fundamentals right, and was very targeted towards art professionals. Let's be honest: the amount of times a normal user would truly use a 13/15" display as a tablet is pretty low, hence the reviews from customers. It was also not a "thin and light" laptop like everything else was trying to be at the time (Apple's 2015 *MacBook* was quite literally 2 pounds). Despite the lack to appeal to mass consumers, it was still one of a kind and truly spoke volumes that Microsoft could be up there and compete with Apple as a design king in the market.

The Evolution of Technology

2015's Microsoft Surface Book

By 2016, the well of the MacBook Air was drying out, the Retina MacBook Pro was now 4 years old, and the 2015 12" MacBook was hit with so much criticism that Apple really needed to come out with a bang. Introduce, the 2016 MacBook Pros. This new generation of MacBook Pros dictated Apple's design language and the features they provided in their computers for quite some time. For starters, the MacBook Pro now came with a function row replacement: the Touch Bar. This mini-display was dubbed as Apple's compensation for a touch screen, and was able to work seamlessly with the computer to provide extra help and buttons for users to take advantage of. There were 2 divides to this: one party that was completely against the entire concept of it and the practicality it took away, and another

Chapter 4: Laptops

which adored it and welcomed the new innovation from Apple.

2016 MacBook Pro (13/15")

It also featured one of the worst keyboards of all time: the Butterfly keyboard. Introduced with the 2015 MacBook — a laptop which truly set so many things backwards for Apple — it was a much more shallower, lack of key travel keyboard that dust would easily get trapped within. However, it was a drastic design change from previous MacBook Pros, as people started to get just a little bored with the silver aluminum chassis.

The Evolution of Technology

12" MacBook's troublesome butterfly keyboard

Microsoft came right back at the new innovations in the laptop market with their first, true laptop with no 2-in-1 features in 2017: the Surface Laptop. Designed to be a mass-marketable laptop, it combined the fundamentals with the phenomenal Surface features we were accustomed to seeing. It featured an Alcantara, soft to the touch keyboard with a 2256 x 1504, 13.5" touchscreen display. Most notably, this laptop features a 3:2 aspect ratio, which consumers and reviewers absolutely loved. It made scrolling and editing through documents much easier, and also gave the visual illusion of a big display. By starting at just 2.76lb with an Intel Core i5 processor, phenomenal battery life and the simplicity of Windows 10, this was *the* computer to buy from Microsoft.

Chapter 4: Laptops

The Alcantara wrapped Surface Laptop

In 2008, when Steve Jobs opened the MacBook Air from an envelope, you would expect this to be the mainstay and flagship laptop of Apple's for at least a decade to come, but it simply didn't. Once the 2012 retina MacBook Pros were introduced, the MacBook Air took a second seat to the more expensive, more premium older brother, and it became extremely dated by its 2015 model. Not every user needs an expensive MacBook Pro however, and it was soon in Apple's best interest to revive the MacBook Air, especially because of the icon and legacy that it held. In 2018, Apple finally brought back and completely redesigned the machine, bringing it much closer to the MacBook Pro models. However, it now started at $1,199, which didn't sit right with consumers as it still lacked some higher level features. This product was definitely not a blockbuster commercial success, but it did *revive* the lineup from Apple, also rendering the 12" MacBook obsolete.

The Evolution of Technology

The resurgence of the MacBook Air.

The year 2020 will be forever known is history as a year where nothing went right. Well, except for the birth of Apple Silicon MacBooks. November 2020 introduced Apple's first processors, and it started the death of Intel processors in MacBooks, and Apple Silicon was simply better. Faster? Check. More efficient? Check. Better battery life? Check. 8-core CPUs meant that these M1 processors were almost just as fast as the 16" MacBook Pros which costed over $3,000! The M1 MacBook Air started at just $999, and it remained with a very similar chassis and minor design upgrades over the previous version, but the performance was simply unparalleled at that price point and category.

Chapter 4: Laptops

Same design, new M1.

The M1 MacBook Pro, the older brother, started at $1,299 at featured an extra fan, increased GPU capabilities, a Touch Bar, faster IO, and better battery life. A bunch of minor upgrades you might think, right? Well, that was the review by critics, as the MacBook Pro was for people who wanted just a little bit more than the Air, with the same exact processor. But overall, the point that still stood was the performance Apple was able to output was second-to-none in the ultrabook market, and put nearly every company on their heels to match Apple.

Apple's Flagship Laptop: The M1 MacBook pro

Fast forward to modern day, and Apple Silicon processors are in every laptop that Apple sells today, and the conversion has been completed with the introduction of the modern 14" and 16" MacBook Pros. Apple unveiled their M1 Pro and M1 Max processors — ones with exponentially superior GPUs to the M1 processors and sizable increases to the CPU's compared to the M1 processors. The design of these new MacBook Pros were ones which looked like a bloated iPhone with the bezel less design and notched display. It quite literally *pushed the edges* on design for Apple, and allowed for much more real estate with the thinned bezels. However, these laptops started at a whopping $1,999 for the baseline 14" MacBook Pro, and users were seeking a more affordable laptop with a similar, futuristic design.

Chapter 4: Laptops

Modern-Day 14" & 16" MacBook Pros

Enter, the 2022 MacBook Air with new M2 processors. These processors weren't quite the upgrade the M1 processors were over the old Intel chips, adding simple CPU and GPU upgrades of around 20% over the previous generation. What was new, however, was the thinner bezels on the display and the notch being introduced as well. It wasn't as elegant and polished as the MacBook Pro 14", but it was Apple's demonstration of bringing the design into the rest of their lineup and engraving that design into MacBook history for future designs.

The Evolution of Technology

Return of MagSafe on the 2022 M2 MacBook Airs

Part Three

Chapter 5: Smartphones – Part One

The definition of a "smartphone" is actually one that *isn't* explicitly defined. Some may consider as a *touchscreen* device with calling abilities. Others may deem it as simply a device that can call, no touchscreen required. For the most part, however, people utilize the former definition. A *touchscreen:* that is what makes a phone *smart* in most peoples eyes.

So, what was the first touchscreen phone, then? Most believe it to be Apple's iPhone in 2007, but there was actually a phone with a touchscreen invented all the way back in 1992, made by IBM. The IBM Simon was a one pound, $899 phone that had a functioning touchscreen, albeit low quality and without good feedback. However, there was an included stylus in the phone (Galaxy Note, anyone?) which made it much easier to navigate the applications and the phone itself. Coming in at a price tag of $899 makes it seem like the phone didn't really break the bank. However, when we adjust for inflation, this phone would have cost roughly $1800 today.

IBM Simon Smartphone

Fast forward 14 years later, and LG introduces the first true smartphone, the LG Prada. This device came with a capacitive touchscreen, making it superior to the IBM Simon introduced back in 1992. Capacitive touchscreens recognize a human touch to the display and mark it down as an electrical charge to notify where the display was truly touched. This made it far superior and more precise as well, and almost every touchscreen device today features this technology. The LG Prada started at a whopping $849, as it was released and named after the Italian luxury designer brand Prada.

Chapter 5: Smartphones - Part One

LG's Prada Designer Phone

Well, well, well. As we all know, Apple's 2007 keynote changed the world forever, and Steve Jobs absolutely revolutionized the way we do things today. The Apple iPhone. Dubbed as *"the first smartphone,"* the iPhone featured a sleek, compact design with the world's most advanced touchscreen panel. Coupled with stock applications, an intuitive display and functionality, and a $499 price tag, this was the phone for *everyone.* Steve Jobs was also light years ahead of the game with this product, as it featured a built in camera, home button, and most importantly, *you didn't need a stylus.* Many touchscreen devices in the past simply wouldn't function well, and a stylus was required. However, not this one.

The Evolution of Technology

Apple iPhone from 2007

However, not everybody was sold. *"How in the world could Apple provide the functionality of a traditional keyboard with a touchscreen"* and *"This is Apple's first ever phone, no way its reliable yet"* were two often complaints that keyboard-truthers had stuck onto. Capitalizing on the still huge market of phones with keyboards, the HTC Dream was introduced. This was truly the best of both worlds — featuring a capacitive touchscreen display, but *also* a keyboard which slid out of the back. The Dream was hit with very high praise from the media and analysts, but there was still one flaw: Android.

Chapter 5: Smartphones - Part One

HTC Dream

Soon after Apple's iPhone, Android was created as an open-source platform as the Operating System for almost every smartphone not called the iPhone. However, it wasn't nearly as refined as Apple's iPhone OS (later changed to iOS), and was met with slight backlash when first released on the HTC Dream.

The Evolution of Technology

Android Logos

Google's acquisition of Android in 2005 is known to be one of the most successful business decisions of all time, and it was time for them to enter in the smartphone race. In January 2010, Google launched their Nexus lineup with the *Nexus One*. However, there was simply a lot of controversy and negative buzz around this phone once it released. For one, there was an entire lawsuit regarding a possible copying of the "Nexus" moniker from a 1968 novel. With first glance, this might seem far fetched, but considering the title of that book was *"Do Androids Dream of Electric Sheep,"* maybe they were on to something. Couple that with 3G connectivity issues, and bugs from the get go, this smartphone simply didn't take off in the market.

Chapter 5: Smartphones – Part One

The Google Nexus One

A couple years later, in 2010, Samsung decided to enter the game of the smartphone with their Samsung Galaxy S. Known for their phones already, Samsung was a veteran in the market aiming to make a splash, and they did. Known even today for their industry leading displays, the 4" Super AMOLED display featured on the Galaxy S in 2010 took the market by storm, and this was simply a refined product with well-functioning Android now on it as well. Selling well over 25 million units as of 2013, this smartphone was one of the most well received and well rounded smartphones of all time. This was a smartphone that also changed the market forever, as the Galaxy S lineup still dominates to this day all across the world.

The Evolution of Technology

Samsung Galaxy S Smartphone

Yes, the iPhone generated shock waves throughout the world in 2007. However, Apple's innovation wasn't done here. Following minor updates to the original iPhone with the 3G and the 3GS, Apple decided they'd perfected the phone part of smartphone, and was time to unleash their monster: the iPhone 4. This iPhone was the walking definition of *perfect.* Featuring a front-facing camera, improved 5MP back camera with flash, squared out design and the introduction of numerous applications for iOS, Steve Jobs had done it again. This phone is known around the world for being most people's first smartphone *ever*, and it was for good reason.

Chapter 5: Smartphones – Part One

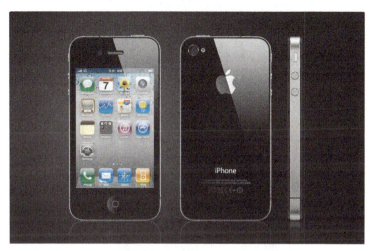

Apple iPhone 4: A Design Beauty

Smartphones and styluses: it was just something that Samsung couldn't let go of, and they channeled this stylus-need through the Galaxy Note of 2011. It featured a 5.3" display which was simply unheard of at the time when smartphones hovered around 4 inches, and it was dubbed by analysts as a *"phablet."* One fan favorite feature it introduced, however, was the S Pen, a stylus that was much more precise and accurate than older styluses and could navigate through the phone seamlessly. The Note lineup has since been synonymous with higher-end features, cutting-edge technology, and *huge* displays.

The Evolution of Technology

The classy, sleek looking Galaxy Note

In a direct contrast with the Samsung Galaxy Note 5.3" display, the iPhone 4S of 2011 still remained with a paltry 3.5" display, and the rest of the market was getting much larger. Enter, the iPhone 5. Released in 2012, Apple focused to make their smartphone on par with the rest of the market in many different aspects, while still retaining their own specialty features. The iPhone 5 came with a much taller 4" display, featured an 8MP back camera and improved front camera. In addition, Apple

Chapter 5: Smartphones - Part One

retained the boxy look and made it *even thinner and lighter* to cater towards the demands of the time.

The two-tone, taller design of the iPhone 5

At this point in time, it was Samsung and Apple for the best smartphone makers, and Samsung didn't sit quiet in 2012. Their Galaxy SIII had a much more homely look, larger display, and a lot more curves than any other phone in the market. This was widely known as the "best phone of 2012" through many analysts, and it was due to the elegance it had gained both software and hardware wise. With an arguably superior camera to the iPhone 5, larger AMOLED display, and much less of Samsung's bloatware on Android, this was truly the iPhone beater of the time.

The Evolution of Technology

Samsung Galaxy SIII

The year of 2013 didn't bring many notable advanced to the Galaxy or iPhone lineup, but one Android smartphone still stuck out of the crowd: the LG G2. This smartphone had much thinner bezels than the competition, featured a 3,000mAh battery to fuel its display, and an overall great package for people who wanted a larger phone, almost a Note competitor. Most importantly, the LG G2 packed a phenomenal camera system, with a 13MP back camera sensor leading the way to its glory. It was also very unorthodox in some of its design features, as the power button and volume keys were on the back of the phone, leaving the sides completely empty. Weird phone? Definitely. Good phone? Yes, as well.

Chapter 5: Smartphones - Part One

LG G2

One of the largest smartphone manufacturers around the world today is OnePlus, and they were a company that wasn't a pioneer in the industry (releasing their first phone in 2014) or had roots deep into the evolution of the smartphone. However, their journey began with the OnePlus One: a phone dedicated to providing as many high-end features as possible for a much lower price point than the competition. This phone had a much more cleaner, close to stock version of Android running on it, with a 5.5" LCD display, and a great battery life. There was no United States version with a supported carrier, which was a major problem for US residents to purchase this $299 phone. This price point was its major selling point, and when offering close to 80% of the same phone as the higher end flagships, it turned eyes across the market.

The Evolution of Technology

OnePlus One: the first OnePlus smartphone

Apple, the face of the industry, were still innovating, and finally catered to the phablet market. With the release of the iPhone 6 and iPhone 6 Plus, Apple pushed the boundaries on design, display, cameras, and size. These phones featured 4.7" and 5.5" displays, respectively, and an improved front and back camera through enhanced software such as auto focus. The boxy look was now gone, however, and the iPhone 6 lineup brought in the curvy design the rest of the market had transitioned to. By catering towards the larger phone market, and still packing features like Touch ID, Apple Pay, thinner design, and light weight, this was still the king of the smartphones.

Chapter 5: Smartphones - Part One

iPhone 6 Plus: a 5.5" beauty

So, did Samsung respond with their own redesign in 2014? Nope. The Galaxy S5 was known as a filler phone throughout the industry, and featured upgrades that people simply weren't pulled towards, and didn't support. However, Samsung decided to go all out for 2015, and made the *best looking smartphone* at that point with the Galaxy S6. Glass, glass, and even more glass. With the S6, S6 Edge, and S6 Edge+, Samsung turned the whole phone into a glass sandwich, back and front. With the introduction of the Edge models, the S6 also had glass melting over the edge for users to utilize during daily life as well. How useful was the glass on the edge? Not that much. Did it create a mesmerizing design never seen before? Definitely. Overall, the S6 stole the show in the smartphone world in 2015 with its brilliant design, and was far more awaited and hyped for in 2015 compared to the incremental iPhone 6S.

The Evolution of Technology

Samsung Galaxy S6 Edge

You might be asking yourself: *"where has Google been all this time?"* With limited success to take over the smartphone industry with their Nexus lineup, Google decided to throw the towel in with Nexus and restart with the *Google Pixel.* Introduced in October of 2016, the Pixel and Pixel XL were phones that were praised for featuring stock Android and its phenomenal software for the camera, but also criticized for its lack of basic features like waterproofing and lack of creativity with the design.

This smartphone was extremely polarizing due to its software, and how Google truly utilized what it was known for to create a great phone, but it simply fell a little behind the market in terms of innovation, design, and hardware. Starting at a price of $649, it had to compete with the likes of the iPhone 7 and Galaxy S7 in order to capture market, but it fell short in doing so.

Google Pixel back view

To wrap up 2016 came the Galaxy Note 7. This was a phenomenal smartphone — it featured the beautiful AMOLED Edge display, came with the S Pen, and even improved upon the Galaxy S7's already great cameras. It's 3500 mAh battery was absolutely HUGE for the industry, and that's where it all went wrong for Samsung.

Samsung Galaxy Note 7 (pictured on the left)

Later in the release cycle, Galaxy Note 7 users reported the batteries overheating and exploding in many different instances, all over the world. Samsung was receiving extreme backlash due to this debacle, and they were required to recall every single Note 7 sold, and customers had to bring it back to their retailers immediately. This caused a 33% decrease in revenue for Samsung, and some users were even scared to purchase another phone from them for quite some time. Their smartphone was plastered through every airport, store, and various other places to make sure that people did not still use or bring that phone anywhere due to safety concerns it posed.

It was simply a disaster for Samsung's reputation it had built up for so many years, a safety hazard, and a very large financial loss for Samsung as well. To come back from this, they truly had to bounce back in a major way with their next phone, in 2017.

Chapter 6: Smartphones – Part Two

2017 might have been the *most* influential year for the smartphone, as we saw revolutionary technologies all across the board from every major company, and redesigns occurred so frequently to keep up with the market as well.

Well then. So how did Samsung bounce back from the Note 7? With their Samsung Galaxy S8. The S8 needed to be different. It needed to feature *everything,* with a risk-aversive model that would win over customers and make them forget the debacle the Note 7 was. However, the Galaxy S8 might've been exactly that, as it pushed the limits to edge: literally. Now featuring a bezel less, even larger AMOLED display, the Galaxy S8 and S8+ were one of the first phones to pioneer the "bezel less" movement, and they were beautiful phones. Featuring an improved back camera (albeit, no dual lenses yet), with 5.8" and 6.2" display sizes, this phone resembled the future, and it dictated it as well.

The Evolution of Technology

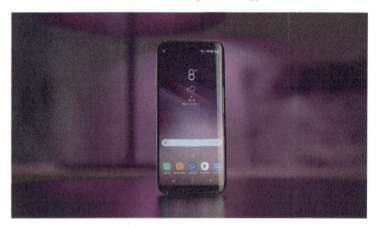

Samsung Galaxy S8: the bezel-less beauty

The LG V30 followed suit with the bezel less movement, as it was known as a fierce competitor in the Android market for people looking for a quality, reliable phone. LG, as most know, is not a brand not as popular like Apple, Samsung, or Google. However, it featured a premium, dual-lens camera with a fingerprint sensor on the back (similar to the S8), with thinned bezels on the front. With a 2:1 aspect ratio and a 6" display, this phone was designed to be the perfect fit for Android lovers, and doing everything the 2017 industry was doing, with nothing really standing out of the crowd. Overall, solid phone, nothing special however.

Chapter 6: Smartphones – Part Two

LG V30 pictured above

Then, in September, came the iPhone X. After a down year for Apple with its iPhone 7 not featuring anything noteworthy, and no main design changes since 2014, Apple made a splash with their iPhone X. The iPhone X featured an OLED display which had close to zero bezel on the sides and the bottom, with only a "notch" on the top. This iPhone featured a dual-lens camera, Face ID, and a 5.8" display: Apple's largest yet. Coming in at a $999 price tag, the iPhone X also set the new price bar for premium smartphones at $999 in comparison to the $649-$749 they were going at a year ago. These prices would only go up from here, and we can all primarily attribute it to the iPhone X, as Apple does dictate the market quite a bit; for example: the notch, the headphone jack, Face ID, etc.

The Evolution of Technology

The "notched" iPhone X of 2017

Following the commercial success of the Pixel 1 and Pixel 2, Google decided they needed to go bigger: and introduce the notch as well. Enter, the Pixel 3 in 2018. The Pixel 3 and Pixel 3 XL featured 5.5" and 6.3" displays, with a notch only being placed on the bigger brother. However, this design and display for the Pixel 3 XL was deemed "out of place," and "an attempt at a notch."

These Pixels still carried the phenomenal camera systems of their predecessors, they still featured the stock-Android that their fan base loved, but it continued to lack in the display and design department, which draws eyes to the phone. Apple and Samsung had made their money for more than a decade now through their designs and build quality, but Google simply couldn't do the same. Unfortunately, this is also where the Pixel lineup fell through the roof a little bit, as customers decided to

Chapter 6: Smartphones - Part Two

render the company's capabilities limited.

Google Pixel 3's Clearly White color

Samsung, a company known for the most innovation and change with their phones, decided to shake it all up with the Galaxy S10 lineup, and introduce the *hole-punch cutout* instead of a full bezel. Samsung also introduced FOUR new Galaxy S10s: the S10E, S10, and S10+, and S10 5G. Mainstream smartphone manufacturers started to do this to diversify their lineup and cater to multiple price points to maximize sales.

With their lineup now offering phones with 5.8", 6.1", 6.4", and 6.7" displays, there was a phone for every person. Samsung also innovated through these phones by introducing up to 3 camera lenses in the back for multiple usages. It also packed some of Samsung's classic gimmicks: for example, PowerShare to send power to another phone through the back of the S10, up to 1TB of storage, and more.

Samsung Galaxy S10's phone-stretching screen

So, we learned about the OnePlus One's fame and how it kicked off one of the top smartphone companies at a lower price point, so where was it 5 years later, in 2019. OnePlus was busy at work with their OnePlus 7 Pro: a phone that costed a little more than a mid-range phone at $669, but featured absolutely anything you could imagine. OnePlus achieved a *completely bezel less* design with this smartphone, as they included a pop-up camera from the top for the front-facing lens, which raised durability concerns at the time, but seemed to stand up well. It had a 90Hz refresh rate display, well ahead of the rest of the market, with a phenomenal OLED panel topping it off in the front.

However, OnePlus still lacked behind larger manufacturers in terms of battery life, waterproofing, wireless charging, and primarily, the cameras. These features customers were accustomed to seeing from their top smartphone manufacturers, making this a tougher buy. Despite it all, it was nowhere close to the $1000+ price tags of the flagship iPhone and Androids, and at $669, there was almost no better phone at that price point.

OnePlus 7 Pro's top-bezelless display

The Evolution of Technology

There comes a point in time where you've pushed all of your boundaries, and smartphone companies started to slowly feel this, Samsung being the one leading the race. So, what next? Foldable displays were something on the rumor mill for quite some time, and achieving this fold ability was something that many companies were trying to perfect. Samsung, being the innovators they were, wanted to be first to this with their Galaxy Fold. Now, the crease in between when the displays were folded was always going to be there due to how the panel works, but it was the lack of elegance which didn't propel this product forward. It's 7.3" display inside was beautiful. But again, it was a first generation product which needed a little more time to get refined.

Samsung Galaxy Fold's foldable display

Huawei was a company that slowly started to turn heads at this point in time, but its growth was throttled due to International restrictions and residents of the United States not able to purchase this phone directly from one of the US carriers. However, the Huawei Mate X was still a very innovative and market setting foldable phone in 2019 known as to be even superior to the Galaxy Fold design wise. It featured an even larger, 8.0" display with the display truly pushing the edge and thin bezels all the way around. Featuring a triple-camera setup, this foldable phone seemed much more cohesive, elegant, and refined than the Galaxy Fold did when it came out.

The polished design of the Huawei mate X

Foldable phones didn't have to expand outwards, however, and Samsung made quick note of that to bring back a design older generation seemed to love: a flip-phone, now with a real touchscreen display. The Galaxy ZFlip, introduced in

The Evolution of Technology

2020, aimed to package a high-end smartphone with premium features, but flippable so that it would be smaller in your pocket.

Now, many critics deemed that there was no use to this flipping design whatsoever, as the only thing you gained was saving space, which most pockets could hold anyways. On the flip side, the simple design of this phone is what appealed to many others. It was a phone that still featured 80-90% of the Galaxy S20 lineups features, but had a cool front facing mini display, was less addictive to use due to this flipping nature and the front display showing notifications, and didn't break the bank like the Fold did.

Samsung Galaxy ZFlip: the smartphone, now foldable again

A company that prioritizes catering to almost every customer like Samsung makes sure they've got their flagships in control as well, and the S22 lineup was one of the best in the market

Chapter 6: Smartphones - Part Two

in 2022. The Galaxy S22, S22 Plus, and S22 Ultra one again catered towards numerous price points and display sizes, with the S22 Ultra being the largest and most feature packed as well. The Galaxy S22 Ultra also rendered the Note lineup dead as it included an S Pen built in. These smartphones came with 3 or more camera lenses, and brilliant low-light photography as well. The adaptive high refresh rate display, coupled with the AMOLED display creates an immersive look, and one of the future for this S22.

Samsung Galaxy S22 Ultra with S Pen

OnePlus over the last 5 years has gained a reputation for making phenomenal phones, and the OnePlus 10T is no slouch in the race. At a price point of just $649, this smartphone featured an adaptive refresh rate display up to 120Hz. In addition, it came with a triple-camera system with much better software tuning than in the past to make it serviceable, yet still not on par with the iPhones of the world. OnePlus also threw in many nifty features to entice customers, whether it be the 150W charging, 16 gigabytes of RAM, or Amp Connect. The sleek design packaged with a hole-punched display led to it being one of the most successful phones of 2022.

The triple camera setup on the OnePlus 10T

Google's smartphone division, the Pixel, fell off of the flagship scene once Google demonstrated year over year they couldn't match the likes of OnePlus, Samsung, and Apple in terms of design and build quality. However, their "budget" lineup, the A-version Pixels, caught steam and were phenomenal phones at their price point. In 2022, the Pixel 6A released at $449, and did everything you would want it to do at a minimal cost. Since Google's cameras were primarily fueled by phenomenal software tuning, they were able to carry this forward to the younger brother, the 6A. Featuring stock Android, long lasting battery life, and a 6.1" OLED display, this phone wasn't weak, and in fact it was powerful and just enough for many customers to purchase.

The Google Pixel 6A

What was Apple up to in 2022? Creating the iPhone 14 and 14 Pro. Since the iPhone X in 2017, Apple really hasn't changed up the design or major features a lot. The iPhone 14 pretty much was a carbon copy with a couple of increased camera features over the iPhone 13, but it still didn't feature an adaptive refresh rate display that some customers wanted to see. The iPhone 14 Pro, however, finally ditched the notch. Featuring the "dynamic island," a pill-shaped cutout to house the iPhone's front-facing technology, the iPhone 14 Pro was slightly different than the 13

Pro display wise. However, the 14 Pro had even *better* cameras than the 13 Pro, and these were camera lenses that still stayed atop the smartphone camera market, and the video shot on them were simply beautiful. Overall, great phone as always, but nothing innovative or crazy new from Apple.

The iPhone 14 Pro

Chapter 7: Smartwatches

When you think of a "smartwatch," does your mind gravitate towards the Apple Watch, Fitbits, and Galaxy Smartwatches of the world? Well, there were smartwatches prior to these which allowed for the innovation of the modern smartwatch.

In 1999, Samsung announced the "SPH-WP10." Yeah, not a fancy name, and sounds more like a model number. This smartwatch was straight out of a movie with the protruding design, numerous buttons, knobs, and switches, and phone-capabilities. However, this smartwatch did have the capability to call due to the placement of the antenna on the left of the phone, and due to the built-in microphone and speaker, you could theoretically replace your phone with this. Later versions of the smartwatch even included a fingerprint sensor, accelerometer, and more high-end features we see in smartwatches today.

The Evolution of Technology

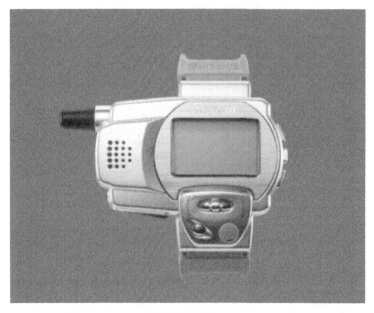

The Samsung SPH-WP10's polarizing design

The year 2001 brought a little more innovation to the smartwatch, and designs closer to resembling today's smartwatches. The IBM WatchPad was created in conjunction with Citizen Watch company, and it had a mini display in the middle of the watch with three buttons for controls and operations right beneath it. With IBM being a main player in the computer segment at this time, they wanted to branch their products out to other devices, yet they simply didn't catch much steam in the market. Even though this smartwatch also had a speaker and microphone to replace your phone, it didn't gain mainstream media attention and didn't sell entirely well too.

Chapter 7: Smartwatches

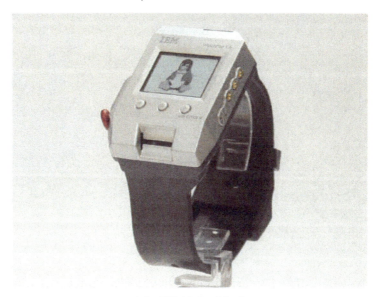

The IBM WatchPad

Fossil was a company that manufactured high-quality smartwatches since the dawn of the 21st century, and their Fossil Wrist PDA was a well refined, well built smartwatch to handle the tasks of the booming generation. Running Palm OS, this smartwatch featured a capable operating system with *runnable applications* in order for higher ranges of productivity. Palm OS's ability to work so seamlessly with touchscreen interfaces made this watch successful. However, the coolest feature about this smartwatch was the built-in stylus it included with the smartwatch, and it assisted users in navigating the display, which was a great addition for users of 2003, a time when styluses were adored.

The Evolution of Technology

The Fossil PDA Smartwatch

Now, all of the smartwatches above *didn't* utilize the old-fashioned circle design like a normal watch, and Microsoft decided to be one of the first in 2004 with the Microsoft Spot. SPOT, which stands for Smart Personal Objects Technology, was an initiative created my Microsoft to create consumer electronics that many people would love and could personalize to themselves. With this smartwatch being able to give you key details of your day (temperature, time, location, weather, as seen below), it was a smartwatch that looked modern, and was different than every other one at the time with its sleek design.

Chapter 7: Smartwatches

Microsoft's SPOT Smartwatch

Fast forward 5 years later, and we are presented with Samsung's S9110 "watch phone." This is one of the more polarizing designs in smartwatch history because it literally looks like a miniature smartphone was strapped on to a watch. However, it created this 1.76" LCD display which users could utilize to call their contacts, and the extra screen real estate meant that more information was visible on the home screens and on applications. Overall, this design probably wouldn't suit well with the smartwatch market of today, but the marketing back then was for it to "replace your phone," which slowly evolved to "assisting your phone."

Samsung's "phone on your wrist"

In 2010, Sony released the Ericsson Live View, a smartwatch which looks very familiar and similar to the Fitbits and Apple Watches of today, a *squircle*. However, this wasn't like your traditional smartwatch, in fact it operated as a Bluetooth display for your Android smartphone. It could transmit live information like Twitter feeds, media playing on your phone, and even more. Despite this, the gist was that it was essentially

Chapter 7: Smartwatches

just a wearable monitor for your phone, and nothing more.

The monitor for your phone: Ericsson Live View

Pebble, a start-up company designed to create innovative smartwatches running their own Pebble OS, released the 2013 Pebble Smartwatch, in what shaped to be a monumental year for smartwatches due to Sony and Samsung's new smartwatches. This smartwatch wasn't flashy, featured a smaller screen than the rest of the crowd, but it was affordable and meant to get non-smartwatch users over to the Pebble world at an affordable price point while still providing key smartwatch features users loved.

The Evolution of Technology

The Pebble Smartwatch

Here is the one you've all been waiting for: the Apple Watch. Released in 2015, this has to be one of the best and most influential smartwatches in the market today. It's design was simply beautiful: Using materials like aluminum, stainless steel, and even a gold version, Apple marketed this smartwatch as "one for everybody," in which different price points existed as well. However, it did feature a smaller display than some competitors on the market, but was won over by the incredible efficiency, speed, and application compatibility. Apple designed a seamless mechanism and way for developers to create and publish apps on the app store, and this propelled into its commercial success.

Chapter 7: Smartwatches

The Apple Watch. Speaks for itself.

On the heels of the Apple Watch every single year, Samsung decided that they needed to do what they did best: design. By creating the Samsung Galaxy Watch in 2018, Samsung wanted one last go at persuading Android *and* iOS users to jump ship to this watch, by making it much more similar to a traditional watch. By featuring a traditional circle display with metal on the edges of the circle watch face, Samsung designed a look that would persuade many to jump ship.

The Evolution of Technology

The Galaxy Watch's "traditional" look

In the late 2010s and 2020s, smartwatches became overly focused on health features, and featuring as many health upgrades they could in their devices. The Fitbit Sense, released in the height of the pandemic in 2020, offered health features never seen before. With a blood-oxygen sensor to track your levels of oxygen, this provided extremely useful to people suffering from COVID-19. They also added a stress monitoring sensor, in addition to the skin temperature sensor (again COVID-19 themed). These features made it *the* smartwatch for people to purchase, but what was the catch? To unlock these features, users had to pay $10 a month for Fitbit's "Premium" plan, on top of the steep $330 price. At the end of the day, most consumers couldn't warrant spending close to $700 if they wore it for 3 years, and didn't purchase this watch.

Chapter 7: Smartwatches

Fitbit's Health-Oriented Sense smartwatch

IV

Part Four

Chapter 8: Smart Home

Smart home devices and automation might seem like they were recently released and invented, and a thing of modern day. However, the roots of smart home technology actually go as far back as the mid-1900s, with there always being a sense of automation desired by technology enthusiasts.

In 1966, Westinghouse engineer Jim Sutherland released the ECHO IV, which critics deem as *"the first smart-home device to be released."* It had the capability to take input, store data, and connect to other appliances throughout the house utilizing this input. The ECHO IV was far beyond its time in that it could control your thermostat, refrigerator, and other appliances! The process of taking in data and storing it was also new to the smart home world at this time, and it allowed for the ability to simply store things like tasks and shopping lists, and then coming back to them later when needed.

The Evolution of Technology

The ECHO IV

However, in 1975, the release of the X10 smart home system was known as more mass-marketable, and was sought after now by more than enthusiasts only. This platform used digital signals and information through radio frequencies onto the direct wiring of a house. Essentially, it fired signals directly onto the electrical wiring of the house's devices and appliances, and controlled them directly through this wiring and through signals sent into this wiring. By using a command console, users of this technology could control their home and various parts inside of it remotely.

Chapter 8: Smart Home

The X10 Smart Home System

In 1990, the Internet of Things (IoT) was born — through a toaster. John Romkey and Simon Hackett invented a toaster that could connect and be controlled through the Internet as part of a challenge they had received. Unknowingly, they created the roots of the Internet of Things, without actually coining the name until 1999. The IoT is utilized today all over the world to control various difference devices through the Internet, hence being able to control from afar. Forgot if you closed your garage door, but are out-of-state? Nothing to worry about, the IoT's got your back.

The Evolution of Technology

The Internet of Things

Despite these early innovations, we see a lot of mediocrity in the sense of revolution, as consumer electronics like laptops and phones gained heavy steam during the late 20th-century. However, in 2005, the Z-Wave platform was introduced, and even to this day, it is one of the most powerful and capable automation platforms to exist. Due to it operating a lower frequency, Wi-Fi and Bluetooth signals don't interfere with the Z-Wave signal, allowing it to communicate freely from device to device. However, the main selling point with Z-Wave was you could convert other electronics to Z-Wave technology with

Chapter 8: Smart Home

just adapters, and this made it extremely versatile.

Z-Wave smart home technology

Now, we talk about commercial success and all, but what is one smart home device that is almost everywhere in households today? The Nest Learning Thermostat. Introduced in 2010, this thermostat made saving energy — and money — much easier for everyone to do. You could control the entire home's temperature, fan, and energy output through a smartphone app connected to the Internet, showing how easy it was to save energy while not in your home. After Google acquired Nest in 2014, it only leaped further into relevancy and at the forefront of smart home devices due Google's presence in the technology and automation sphere.

The Evolution of Technology

Nest Thermostat in the House

In 2014, we also were introduced to Apple's HomeKit, which was their attempt at the smart home industry. HomeKit was platform that Apple created in which users could control their smart home devices all through one app. With Apple's neat software and elegant design, many users were persuaded. Apple's influence in the industry also propelled manufacturers of these smart home devices to be HomeKit compatible, as so many users were in the Apple ecosystem already.

Chapter 8: Smart Home

Apple's HomeKit

2014 brought us Amazon's version of the ideal voice assistant, *Alexa*. Alexa was Amazon's stab at the voice assistant market, but it was featured towards integrating in your home particularly, and not on your phone and other electronics like other assistants. Alexa's special abilities were primarily when connected to other devices, and when paired to devices that had smart home compatibility. At the time, this was the voice assistant that supported the most smart home devices (lights, outlets, fans, and more), propelling it to the forefront of smart home voice assistants.

The Evolution of Technology

The Alexa Voice Assistant

Adding to the monumental year 2014 was for the automation and smart home world, Amazon treated us to the invention of the Amazon Echo. The Echo was a smart-home speaker that featured *Alexa,* the voice assistant mentioned earlier. When pairing Alexa with a speaker that could pick up voices and hear from long distances, Amazon created more than just automation around the house, but took advantage of the hardware the Echo possessed in order to create the ultimate gadget. Want to play some music for your party? Check. Want to write down your shopping list for the supermarket? Check. Trying to dim the lights in your living room while watching a movie? Check, again.

Chapter 8: Smart Home

Once Amazon flew off to instant success with the Echo and its affordable price to be placed in every single household, Google knew it was their time to fire back with their own magic. Let's start off with *Google Assistant*. Introduced in 2016, Google focused a lot more on its Home being adaptable and versatile to be placed on numerous different devices, whether it be a handheld electronic or a smart home speaker. Google Assistant was much more natural than Alexa or Siri, and came with more sophisticated responses and algorithms created by Google to truly emulate the feeling of talking to a real person. However, Googles attempt at rolling out a "one size fits for all" type assistant inhibited its success a little bit in both the consumer electronics and the smart home market, but it nonetheless demanded a high share.

Hi, how can I help?

The Google Assistant

In the same year, as part of Google's plan to compete toe-to-toe with the Echo, the Google Home was launched. This was a smart speaker that aimed to be more "personable" than any other speaker in the market, feeding off of the Google Assistant's primary feature. With Amazon touting all of the products its Echo was compatible with, Google responded with an array of Google Home compatible products as well, bridging the gap that people envisioned was between the two devices. With over 52 Million units sold of the original Google Home, it was a massive commercial success and cemented Google's name in the smart home industry ever since. However, the war was now on between the Amazon Echo and the Google Home.

Chapter 8: Smart Home

The Google Home

Chapter 9: Autonomous Vehicles

To understand autonomous vehicles, one must first understand autonomy and what it encompasses. So, how do autonomous vehicles work, and what are the ranges of autonomy that exist?

Inside a Self-Driving System

Chapter 9: Autonomous Vehicles

Globally, there are 6 levels of autonomy that are defined, and are widely used to describe automobiles.

Level 0: This is the most basic level of autonomy, and it actually means that there's no autonomy. The driver must brake, steer, and accelerate, performing all actions of driving the vehicle with no other assistance. Things like emergency braking systems do *not* count as autonomy, as they don't drive or assist your driving.

Level 1: This is the lowest level with *actual autonomy*. These vehicles have a single automated system that assists the driver. For example, adaptive cruise control would be classified as a single automated system assisting the driver.

Level 2: This is more of the level that a lot of cars we see today are at, and these vehicles have advanced driver assistance systems. The vehicle can both accelerate/brake, and also turn at the same time, but the driver can take over at any point. Tesla's *Autopilot* is an example of a Level 2 autonomous system.

Level 3: The step-up from Level 2 to Level 3 includes a lot more hardware, and a smaller feature-set increase. Level 3 vehicles, like the rumored *Audi A8L with Traffic Jam Pilot*, have Lidar scanners fitted onto the car to enable higher levels of environmental detection. However, human presence is needed and they need to be able to take over the wheel at any time.

Level 4: This is the point where the human presence is not required in almost every circumstance possible. When something goes wrong, and there is an unknown/peculiar situation on the road, Level 4 vehicles *will* be able to make the decisions and work through it. *Waymo* released Level 4 taxi service in Arizona city roads which don't require a driver, and it's been running since 2017.

Level 5: These vehicles don't require human intervention whatsoever, and humans won't have the capability to intervene if they wanted to. With the omission of gas/brake pedals and the steering wheel, these cars are fully autonomous. However, they are currently a thing of the future, and no system is yet to be released with Level 5 autonomy.

nuTonomy - MIT Start-Up for self driving vehicles

Now, let's examine a couple autonomous vehicles over the years that have featured some level of autonomy.

Chapter 9: Autonomous Vehicles

Tesla's *Autopilot* system — one of the most widely used and popular self-driving systems — is found on every new Tesla sold today, the Model S, 3, X, and Y. Featuring Level 2 autonomy, these vehicles have the capability to speed and brake while turning on the highways, and have limited local road functionality yet. However, Elon Musk's ingeniousness has blessed us with a plethora of other features additional that make Autopilot so unique.

For example, Auto Park is a feature where the driver simply rolls up in front of a spot they'd like to park in, and press *Start*, initiating the process to park in that spot. Smart Summon, a feature that's still working out its bugs, can "come to you" in anywhere of the range of 100 feet, or a location that you specify it to go. On a rainy day inside the grocery store, it's an advantageous feature for the car to come straight to the entrance, instead of you having to walk to your car.

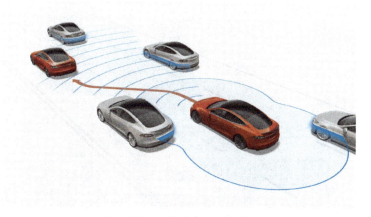

Graphic on Tesla's Autopilot

The Cadillac Escalade is another car featuring some sort of self-driving system, which Cadillac called *Super Cruise.* In a nutshell, Cadillac went around the United States and mapped out over 200,000 miles worth over highways to create a visual map for the Super Cruise to utilize. Prior to the launch of Super Cruise, cars went around mapping the highways systems across the country with high-technology LiDar sensors in order to get accurate data on these highways, and how to maneuver them.

Based on this data, the Super Cruise program allows for drivers to get much more precise tweaks to the system and is much more smoother across the edges than Tesla's autopilot is. In the world of programming, one would generally call this "hard-coded," and it is 100% true. It will never be as impressive as Tesla's autopilot system primarily due to that it isn't processing a lot of this information in mealtime, and is instead relying on previous data.

Cadillac Escalade

Chapter 9: Autonomous Vehicles

Audi, the luxury German car company, had planned to roll out with Level 3 autonomy on their Audi A8, which was in research and production for quite some time. However, the step up from Level 2 to 3 is one of magnitude, and in 2020, Audi decided to scrap this plan for Level 3 autonomy.

What was originally planned as a self-driving system that could steer, accelerate, and brake on highways, in addition to not as much attention needed by a human driver. In Tesla's autopilot, the driver must be active and paying attention, and human signal is necessary as you must apply a *"turning force"* on the wheel ever now and then to show alertness in the car.

Audi A8: The car that could've been great.

The world of the autonomous vehicles is one that is still very young, and there is a lot to come. Whether it be the Waymo taxis running in Arizona local roads at Level 4 autonomy, or

Tesla's mass produced Level 2 Autopilot system, there is a lot of innovation and invention currently being taken place in the automobile world, and it will only accelerate faster from here.

Chapter 10: A Trip to the Future

Over the last 2 centuries, the human race has been expanding at an exponential rate, and with that came the innovations and inventions we use today. Every second, every minute, and every day brings the creation of newer technology, and that begs the question: what could we see in the future?

Drones have been utilized in many different senses already today, with the military being a major contributor to drone sales, but what if we could do more? Amazon is already starting up *shipping packages* with drones straight from their factory, shortening times and removing the need for packages to exchange numerous hands. Drones are also supposedly being created to deliver supplies and resources all around the world for remote areas and people in need of them. Aircraft is something we've mastered since the 20th century, and bringing it down to a smaller scale through drones is something definitely achievable, and could happen very soon.

The Evolution of Technology

Amazon Prime Air Drone

Let's get a little more futuristic now, shall we? The *Metaverse*, still in its early stages with massive plans incoming, it aims to almost replace the real world with a virtual one. What might

Chapter 10: A Trip to the Future

seem like straight out of a movie is actually now coming to real life, and it might be both scary and cool at the same time. The idea and concept of "social connection" through a virtual-reality headset is intuitive and cool to take part of on your free time, which is what currently is going on with users of the Metaverse. By creating mini worlds, social "gatherings," and essentially an entire life with separate friends, pets, and family inside of the virtual world itself, you are now *truly* detached from the world.

A festival in the Metaverse

Have you ever been to a big city, and the traffic is nearly impossible to go through, with it feeling like could you ride a bike faster than the traffic? Well, to resolve those problems, Elon Musk, the famous/infamous founder of Tesla and SpaceX is creating the *Hyperloop* through The Boring Company. This project aims to create a tunnel system all around the country (when finished), in order to create a simple way for people to hop in their car, and allow for the Hyperloop to transport them, with no driving necessary. Allowing for essentially zero traffic and extremely high speeds due to this lack of risk, this is something which would make traveling to far locations and your daily commute much quicker and efficient for the world as well.

Elon Musk's Hyperloop

Artificial Intelligence seems like it's always improving, with Google Duplex AI now even being able to book your haircut appointment for you based on your schedule and availability through Google's other services. However, we are not even close to the capabilities possible through Artificial Intelligence. Automation is a field with endless possibilities to streamline workflow and daily life, with technologies such as detecting the types of clothes you purchase through pictures of your outfits, and then mapping them to stores around to select and

Chapter 10: A Trip to the Future

shop based on your style and preferences available. Artificial Intelligence simply has no bounds or limits, and a lot can be done in the next century to advance the human race.

Technology has truly changed our lives over the past years. Whether it be the streamlined integration of technology into the workspace, the classroom, and society. Almost anything can be defined as technology — even the wheel — and the rate at which resources and money have been spent to advance our society and the technology in it is extremely remarkable.

I want to leave you all with one note: step back and take a second to think of every piece of technology you use in a given day. *Appreciate this evolution, and encourage the continuation of innovation.*

Made in the USA
Columbia, SC
29 October 2022